HIGH TIDE, LOW TIDE

NATURE'S CYCLES

Jason Cooper

Rourke
Publishing LLC
Vero Beach, Florida 32964

www.rourkepublishing.com

PHOTO CREDITS: All Photographs © Lynn M. Stone

Editor: Robert Stengard-Olliges

Cover and interior design by Nicola Stratford

Library of Congress Cataloging-in-Publication Data

Cooper, Jason.
 High tide, low tide / Jason Cooper.
     p. cm. -- (Nature's cycle)
  ISBN 1-60044-178-5 (hardcover)
  ISBN 1-59515-540-6 (softcover)
 1. Tides--Juvenile literature. 2. Seashore--Juvenile literature. 3. Gravity--Juvenile literature. I. Title. II. Series: Cooper, Jason. Nature's cycle.
  GC302.S76 2007
  551.46'4--dc22   2006013283

Printed in the USA

CG/CG

Rourke Publishing

www.rourkepublishing.com – sales@rourkepublishing.com
Post Office Box 3328, Vero Beach, FL 32964

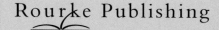

Table of Contents

Changes at the Seashore

A **seashore** changes every day. It may be covered with water. Sometimes sea water rises high onto a **beach**.

But the same shore may be dry at another time of day!
These changes in water level are from changes in tides.

High and Low Tides

Tides are rises and drops in oceans and other large bodies of water. **Tidal** changes are easiest to see along seashores.

High tide carries water higher onto land. Low tides lower the level of water.

Tides are not the same each day. Some low tides, for example, are lower than others are.

The greatest tide changes are in the Bay of Fundy in Canada. High tide is about 39 feet (12 meters) higher than low tide!

Some seashores have one high tide and one low tide each day. Other places have as many as two high and two low tides each day.

Gravity

A powerful force called **gravity** causes tides. We cannot see gravity. But we can see what gravity does with ocean water.

Gravity is a strong pulling force. The Earth, sun, and moon all have gravity.

The gravity forces of the Earth, sun, and moon **tug** against each other. Other forces, however, keep them apart.

Gravity and Tides

The gravities of the sun and moon are strong enough to pull water upward. The tug of their gravity is enough to cause tides on Earth.

The moon is much smaller than the sun. The moon, though, is much closer to Earth. The moon's gravity is more of a tidal force.

The Earth moves through space. Its does not always line up the same way with the sun and moon.

Earth's movement changes the strength of the moon's and sun's gravity toward Earth. Changes in gravity cause changes in tide.

Glossary

beach (BEECH) — a strip of sand or pebbles where land meets water

gravity (GRAV uh tee) — a force that pulls things toward the surface of the Earth and keeps them from floating into space

seashore (SEE shor) — the sand or rocky land next to the sea

tidal (TIDE uhl) — related to the tides

tug (TUHG) — to pull hard

INDEX

FURTHER READING

Herman, G. *Creeping Tide*. Kane Press, 2003.
Jacobs, Marian B. *Why Does the Ocean Have Tides?* Powerkids Press, 2003.

WEBSITES TO VISIT

http://www.enchantedlearning.com/subjects/ocean/Tides.shtml

ABOUT THE AUTHOR

Jason Cooper has written many children's books for Rourke Publishing about a variety of topics. Cooper travels widely to gather information for his books.